Farm Animals

Chickens

Heather C. Hudak

Weigl Publishers Inc.

Published by Weigl Publishers Inc.
350 5th Avenue, Suite 3304, PMB 6G
New York, NY 10118-0069
Website: www.weigl.com

Library of Congress Cataloging-in-Publication Data

Hudak, Heather C., 1975-
Chickens / Heather C. Hudak.
p. cm. -- (Farm animals)
Includes index.
ISBN 1-59036-423-6 (hard cover : alk. paper) -- ISBN 1-59036-430-9 (soft cover : alk. paper)
1. Chickens--Juvenile literature. I. Title.
SF487.5.H83 2006
636.5--dc22
2005034668

Printed in the United States of America
1 2 3 4 5 6 7 8 9 0 10 09 08 07 06

Editor Frances Purslow
Design and Layout Terry Paulhus

Cover: There are more chickens on Earth than people.

Contents

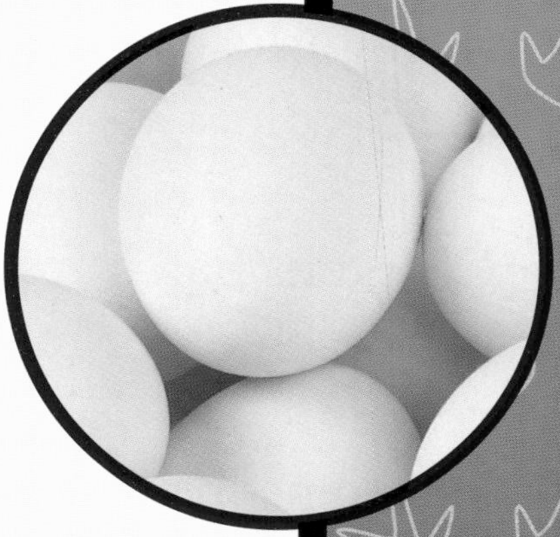

Meet the Chicken

Chickens are birds. A chicken has a **comb** on top of its head and two **wattles** under its chin. Most chickens have four-toed feet with sharp claws. Their feet help them scratch the dirt so they can find food.

Farm chickens have large, heavy bodies and short wings. Most of them cannot fly. Chickens come in many colors and sizes.

Chickens hatch from eggs. Female chickens, or hens, help each other raise their babies.

Chickens live in large groups called flocks. They have a pecking order in these flocks. In a pecking order, certain chickens lead the group. Chickens at the top of the pecking order are the first to access food and nests.

Fascinating Facts

Before 1930, chickens were used for their eggs more often than for their meat. However, as more people bought refrigerators for their homes, farms produced more chickens for meat.

Heavy breeds of chickens can flap their wings and jump. Lighter chickens can fly short distances.

All about Chickens

Farmers raise chickens for two purposes. They keep some chickens for their eggs. They keep others for their meat. Chickens raised for meat and eggs are called poultry. Some people keep chickens as pets or to display at fairs and contests.

There are many chicken **breeds** and hundreds of **varieties**. Some breeds are named for the part of the world they come from. Others are named for their use.

Chickens are social animals. They can recognize each other by their facial features.

Breeds of Chickens

American	Asian	Silkie Bantam

• First came from Europe • Varieties include Rhode Island Red and White Leghorn • Used for meat	• First came from China • Varieties include Brahma, Cochin, and Langsham • Used for meat	• First came from Indonesia • Varieties include Black, Blue, Partridge, Buff, White, and Gray • Used for display

English	French	Mediterranean

• First came from Europe • Varieties include Sussex and Dorkings • Used for meat or eggs	• First came from Europe • Varieties include Faverolles and La Fleche • Used for meat and eggs	• First came from Europe • Varieties include Ancona, Minorca, and Leghorn • Used for eggs

Chicken History

Chickens were some of the world's first tamed animals. The first chickens came from southeast Asia. They were called red jungle fowl. These fowl were raised in Vietnam and Thailand about 8,000 years ago. All chicken breeds today are related to the red jungle fowl.

Chickens were first brought to China in 1400 BC. As well, chickens are mentioned in Greek books and poetry from 400 BC.

Phoenix chickens were brought to the United States and Great Britain in the late 1800s.

Fascinating Facts

Babylonians carved pictures of chickens in 600 BC.

China has more chickens than any other country. About 3 billion chickens live there.

Chicken Coops

On farms, chickens live in shelters called coops. Coops protect the birds from injury, poor weather, and **predators**. Chicken predators include coyotes, foxes, raccoons, hawks, and opossums.

Chickens need to have a **run** outside the coop. The run must be fenced and covered to keep out predators. It is a good idea to have a dog to guard the run and coop.

Chickens need **perches**, nesting boxes, food, and water inside the coop and the run. There should be a source of light inside the coop year-round. Coops also need windows to let in fresh air. Clean, dry straw or wood shavings make good bedding for chickens.

Chickens make good pets because they are easy to care for and train.

Chickens need to leave the coop so they can hunt insects, roam, and peck. Wire fencing stops them from wandering too far.

Chicken Features

Chickens are **adapted** for living on the ground. This is where they find their foods, such as worms, insects, and seeds. Chicken feet are designed for scratching the earth. Wattles and combs decorate the head of chickens. The wattle and comb help keep chickens cool in hot weather. Male chickens have larger combs than female chickens. Combs come in many shapes and sizes, depending on the breed.

FEET
Chickens have four-toed feet. They use their feet to scratch at soil and dirt.

FEATHERS
Chicken feathers, called plumage, can be many colors. Feathers can be white, gray, yellow, blue, red, brown, or black.

NECK
Chickens have twice as many neck bones as humans.

HEAD
The combs of most chickens are red.

What Do Chickens Eat?

Chickens are omnivores. Omnivores eat both plants and animals. Chickens eat fruit, herbs, leaves, acorns, grains, and seeds. They also eat insects, worms, slugs, and snails.

Chickens raised on farms eat special feed that contains **protein** to keep them healthy. Some chickens get their protein by eating small mammals, such as mice.

When chickens are laying eggs, they need extra **calcium** in their diet. Oyster shells are a good source of calcium for chickens.

Chickens have a gizzard. This is an organ behind the stomach. The gizzard has tiny stones that grind up food. This helps chickens digest their food.

Fascinating Facts

Baby chickens, or chicks, can go without food for one week after hatching. They feed off yolk that is in their stomach.

Chickens search for insects in the dirt by scratching with their claws and pecking with their beaks.

Chicken Life Cycle

Male chickens are called roosters. Female chickens are called hens. Baby chickens are called chicks.

Hens begin laying eggs between 18 to 20 weeks of age. They can lay one egg each day for up to 30 days.

Egg

An egg begins as yolk inside a hen. When the egg is fully developed, the hen lays the egg. A hen will sit on the egg for 21 days to keep it warm. Hens also turn their eggs regularly with their beaks.

Chick

When they are ready to hatch, chicks peck their way out of the eggshell. They make a peeping sound. Hens hear this sound and cluck to urge the chick to break through the shell. Chicks do not have full-grown feathers. Instead, they are covered with fluffy, yellow **down**. Chicks can walk as soon as they hatch. They can also feed themselves.

On farms, hens sometimes try to lay eggs in nests that already have eggs. Hens also try to move eggs from other nests into their own nests.

Some farmers use fake eggs to encourage hens to lay in particular locations.

Adult

Chickens are full-grown at 32 weeks of age. Adult chickens can weigh between 1.25 and 12 pounds (0.5 and 9 kilograms), depending on the breed. Chickens often live up to 10 years. With special care, some chickens can live as long as 22 years.

Caring for Chickens

Chickens need special care. Farm chickens are often de-beaked. This means that part of their beak is removed. De-beaking stops chickens from pecking at wounds on their skin or other chickens.

Sick chickens should see a veterinarian, or animal doctor. Chickens can become sick with many types of **diseases**. These diseases might include scaly leg, fowl pox, and colds. Chickens can also get **parasites**, such as lice, mites, fleas, ticks, and worms.

Eggs must be turned over three to five times each day. If they are not, the chicks inside will stick to one side of the shell. Hens will do this. If they do not, farmers will turn the eggs for them.

Useful Websites

To learn more about raising chickens, visit:
www.birdhobbyist.com
and click on the care sheet.

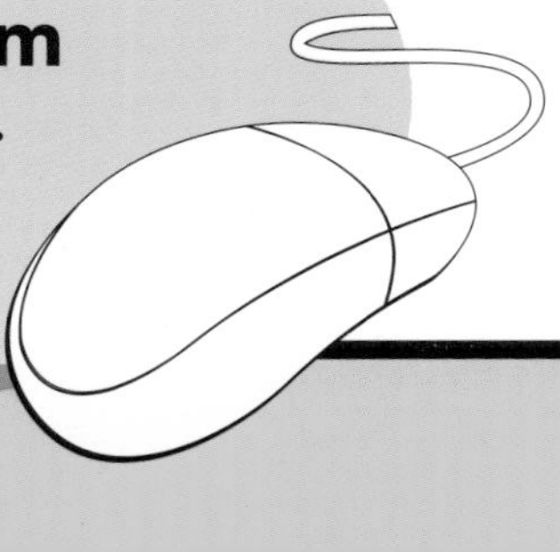

Chickens need special care and attention. Raising chickens means feeding, watering, and washing them.

Myths and Legends

Some cultures have special beliefs about chickens. In Indonesia, some people bring a chicken to funeral services. They believe evil spirits will travel into the chicken rather than the people gathered at the service.

Ancient Romans thought that the behavior of chickens could **foretell** the future. Ancient Greeks believed that chickens were brave and fearless. They even thought that lions were afraid of chickens.

Chinese peoples have a calendar that links each year with a certain animal. The rooster is one of the animals on the Chinese calendar.

Those born in the year of the rooster are thought to be good thinkers, talented, and brave.

Half-Chicken

Author Alma Flor Ada retells a Mexican folktale about a special rooster that became known as the weather vane.

A long time ago on a Mexican ranch, there was a chick that had only one wing, one leg, one eye, and half as many feathers as the other chicks. He was called Half-Chicken.

One day, Half-Chicken decided to go to Mexico City. Once there, he found himself at the kitchen door of the palace. The cook threw him into a kettle of water. Half-Chicken cried for help.

The cook left and when he returned, he saw that the water had spilled and the fire was out. The cook then picked Half-Chicken up by his leg and threw him out the window. The wind blew fiercely, lifting Half-Chicken higher and higher, until the little rooster landed on one of the towers of the palace.

From that day on, weather vanes in the shape of chickens have stood on one leg pointing into the wind.

Frequently Asked Questions

Why are chickens sometimes called fowl?

Answer: At one time, fowl was the name given to any kind of bird. Today, it most often refers to chickens that are raised on farms as pets or as food.

What are broody hens?

Answer: Hens that stop laying eggs are broody. Broody hens sit on the nest until their eggs hatch. They seldom leave the nest even to eat or drink.

Why do roosters crow?

Answer: Roosters like to have their own space. They crow to warn other roosters to stay away. Roosters begin every day by crowing.

Puzzler

See if you can answer these questions about chickens.

1. What are three chicken breeds?
2. Where did chickens first come from?
3. What are chickens used for?
4. What do chickens eat?
5. How long do hens sit on their eggs?

Answers: 1. Asian, American, Bantam, English, French, Mediterranean 2. Southeast Asia 3. Meat, eggs, pets 4. Insects, worms, fruit, herbs, leaves, seeds, acorns, slugs, and snails 5. 21 days

Find Out More

There are many more interesting facts to learn about chickens. If you would like to learn more, take a look at these books.

Sklansky, Amy. *Where Do Chicks Come From?* New York: HarperCollins, 2005.

Gibbons, Gail. *Chicks & Chickens*. New York: North Adams, Holiday House Inc., 2003.

Words to Know

adapted: adjusted to the natural environment

Babylonians: people from Babylonia, an ancient region in Europe

breeds: groups of animals that have common features

calcium: an element that helps form bones and teeth

comb: a ridge on the top of a chicken's head

diseases: serious illnesses

down: soft, fluffy feathers on young birds

foretell: a signal that something will happen in the future

parasites: living things that feed off and live on other living things

perches: places to sit or rest

predators: animals that hunt other animals for food

protein: a nutrient found in meat, eggs, and nuts

run: an animal enclosure

varieties: groups that share certain features

wattles: colorful, wrinkly folds of skin that hang under the chin

Index